Mobile Solar Power:

DIY Installation Mobile 12 Volt Off Grid Solar System

With Step-By-Step Instructions

Disclaimer: All photos used in this book, including the cover photo were made available under a Attribution-Non Commercial-Share Alike 2.0 Generic and sourced from Flickr

Table of content:

Introduction: Understanding Electricity and Solar Power 4
    What is Electricity? 4
    How is Electricity Measured? 6
    Electricity through Solar Power 7

Chapter 1: What Solar Power Systems Consist Of 8
    Solar Panels 8
    Solar Cells 9
    DC Wiring 10
    Deep Cycle Batteries 11
    DC/AC Inverter 12
    Mounting Equipment 13
    Utility Meter 14
    Generator 15

Chapter 2: Mobile Solar Power Setups 16
    Mobile Mason Solar Powered Light 16
    12 Volt Portable Solar Power Bank 17
    Solar Power Mint Tin 19
    Pepsi Can Solar Panel 20

Chapter 3: Monitoring Your Power Bank Voltage 22
    Calculate Just How Much Power You Need 22
    Know How to Log and Maintain Data 23
    Don't Fry Your Battery 24

Chapter 4: Further Maintenance 25
    Regularly Inspect Your Panels 25
    Keep Your Solar Panels Clean 26
    Keep Track of Sun and Shade 27
    Monitor Overall Usage 28

Conclusion: Mobile Power on the Go! 29

Introduction: Understanding Electricity and Solar Power

The first step of creating your own DIY mobile solar power system, is to develop an understanding of electricity and solar power for yourself. So, let this introduction guide you through the entire process that occurs when you plug in your favorite device and power it up.

What is Electricity?

At its most basic level, electricity is simply the movement of charged particles. These particles can be found in nature through lightning bolts, and even in our own bodies by way of the thousands of nerve impulses that send electrical signals sent from our brain and to our hands, arms, legs, and a whole host of other aspects of our physical form. If electricity then is simply the "flow of electric charge" where does this charge come from? Is it simply the spark of life itself?

Well—perhaps in some bygone dark age we would have chalked up the phenomenon of electrical charge to some mysterious unknown force.

But we now know where indeed that this charge came from—they come from the atomic building blocks of matter itself; protons, neutrons, and electrons. These fundamental elements work on a positive and negative charge. And once this charge is created it can do one of three things; it can "produce" electricity, it can "store" electricity", or it can "consume" electricity.

A mobile solar power unit for example, produces electricity, whereas batteries are primarily used for storing electricity and all manner of our devices—such as this battery powered laptop I'm typing this on right now—consumer electricity. And when we receive the electricity that we use to power our devices it comes to us in what is known as DC power, which is an abbreviation for "Direct Current". For most appliances however, DC just won't do, since most circuits are created to run an "AC" or "Alternating Current".

AC is used simply for the sake of efficiency, since an alternating current can travel much farther than a direct current can. It is for this reason all electricity created from solar power in DC form must be run through an inverter to change it to AC. We will discuss inverters and the process of converting DC to AC a little further on in this book.

How is Electricity Measured?

The three main vehicles used to measure electricity are in "Volts", "Amps", and "Watts". Volts measure the full potential of the electrical power that goes through a wire or any other conducting element. These volts can be measured and quantified beginning with the number 1—with calculations such as 2 volts, 5 volts, 9 volts, 12 volts, and so on, and so forth.

Amps on the other hand, measure the exact amount within the current at any given time. Amps are only quantified when they are being actively pushed through a connection to a device. But it is watts that measure the total amount of power that is created by adding the total volts and amps of an electrical current together. Watts, Volts, and amps are all used when it comes to measuring electricity in general, and in this book, it will be use to quantify power when it comes to solar power in particular.

Electricity through Solar Power

Solar power has been used in one way or another for the past few thousand years. Before the washer and dryer, most washed their clothes in the river and then dried them stretched out in the sun. Whole civilizations such as those found in ancient Egypt, also used the sun for solar powered heating. With their rudimentary understanding of the solar process, they purposefully designed buildings to absorb as much soon as possible during the day, and then empty out the heat in the cool of the night.

After the Egyptians, the Romans were the next to extensively utilize solar power in the form of their solar heated bath houses. These ancient saunas were used by just about everyone in Roman society for health, wellness, and relaxation. And now in the modern era methods of solar collection have been perfected to a science. Solar energy is absorbed through the use of photovoltaic solar cells. These solar cells are placed within a special housing called a "solar panel". It is these solar panels that you often see hoisted up above houses to collect solar power. We will learn all about these solar power systems in the next few chapters in this book.

Chapter 1: What Solar Power Systems Consist Of

There are quite a few solar power systems that you can use to collect solar energy, but here in this chapter we will run a basic overview of the main components that most consist of. Here are a few examples.

Solar Panels

The solar panel itself, is actually the framework or housing that solar cells go into. They are usually composed of light wood or aluminum materials, shaped into a rectangle. For the purpose of our tutorial in this book we are going to use a ½ inch thick piece of ply wood with a raised lip of about 1-inch thickness on its outer edges. This lip will serve to hold the solar cells in place on the panel.

Inside this plywood frame you then need to get some pegboard. Pegboard, with its perforated, and perfectly placed holes, serves as the perfect backdrop for your solar cells. Just take your pegboard material and cut it down to the same size as the inside of your solar panel frame, and then use wood working glue to fix it in place, inside your solar panel. Once you have done this, consider your solar panel complete, all you need to do now is attach your solar cells.

Solar Cells

Arguably the most important aspect of the entire solar system, solar cells are the component that absorbs direct sunlight. You can buy these cells at stores or online. But just a word of caution—if you plan on buying these cells new you are liable to pay a pretty penny for them. But solar cells purchased used or slightly blemished on the other hand, can be bought for literally just a few bucks a piece. Fortunately, there are used solar cell stores, sites, and distributors popping up all over the place.

And although these cells are used or slightly blemished, it does not affect their solar absorbing capacity in the slightest. Usually the blemishes are just simply nicks, cuts, or scrapes across the surface that do not directly affect the solar cells ability to collect sunlight. These are just minor cosmetic defects caused from either error in manufacturing or through accidents in handling. Solar cells cannot be sold by expensive retailers unless they are in pristine condition.

So, if the delivery truck they were packed in hits a bump in the road that causes a box to fall and crack a few solar cells, those cells can no longer be sold as brand new and become blemished. These are then handed off to distributors who sell these slightly dinged up cells for a much cheaper price. These cells may not always look as aesthetically pleasing as their streamlined, unblemished brethren, but they still get the job done. Even with a big crack straight down the middle, these cells usually have no interference in collecting solar power. They might look a little banged up but they still collect plenty of solar power.

DC Wiring

Once your solar panel and solar cells are in place, you are going to have to think about the kind of DC wiring you are going to install. For the most part, DC wiring is pretty simplistic and straight forward, and just about anyone could do it. All you need is a power-drill, a set of sunlight resistant solar panel wiring, and a special device called a "blocking diode" that will keep the battery your panels will be connected to from becoming drained during periods of inactivity.

The power-drill I believe, is a self-explanatory item and can be purchased at just about any hardware store. As for the solar panel wiring itself, these are typically made out of aluminum or copper material, with copper being perhaps more expensive but by far the more efficient choice when it comes to solar power. Copper facilitates the flow of solar energy much better than aluminum does, and is much more durable than aluminum-based alloys.

After drilling a hole into the back of the frame of your solar panel, you are going to run your solar panel wiring through it, and solder the end point direct to your solar cells. Next, run the wire through a blocking diode. Now take the termination point of the wire and hook it up to a "polarized two pin" plug and hook it up to a "deep cycle" battery. We will discuss deep cycle batteries more in depth in the next section.

Deep Cycle Batteries

Deep cycle batteries are the storage units for all of the solar energy that you collect direct from the sun. These batteries work as your solar bank, so that even on the gloomiest, most overcast of days, you can tap into it for energy you stored up when the sun was shining bright overhead. These batteries were made with storage of power in mind, holding onto energy for long periods of time and then deeply discharge that energy when you need it, sending it right over to whatever device or apparatus you need to power.

Other batteries, such as the kind you might find in your motor vehicle, are not capable of storing such power for the long-term. Batteries for most modern cars are only able to store very small amounts of electrical power, just enough to get your car started, and then your car's alternator takes over as the true powerhouse of the vehicle. Deep cycle batteries on the other hand, can carry a huge amount of power with them and then are capable of discharging as much as 80% of their stored energy all at one time.

Among deep cycle batteries there are generally three options available; there are "lead acid batteries", "flooded batteries", "gel batteries, and "absorbed glass mat (AGM batteries)". Flooded batteries are popular because they are cheap, but they also emit a lot of fumes during prolonged use, and its for this reason that they need to be kept outside. But outside storage has its own issues, since it can lead to frozen and dead batteries during the winter months. Lead acid batteries on the other hand, are just a little bit more expensive than the flooded variety and can produce much the same results without fear of fumes.

DC/AC Inverter

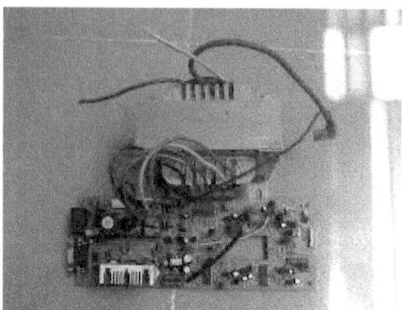

As mentioned earlier in this book, solar power is collected as a direct (DC) current. In order to power your AC (alternating current) driven devices therefore, you will need to run it through an inverter. The inverter will multiple your DC creating 120 AC volts for every 12 DC volts. These kinds of inverters are easily recognizable due to their "cigarette lighter" port and their 120-volt socket.

Mounting Equipment

The mounting equipment for your solar system is the intractable foundation on which you place your solar panels, this is the component that allows solar panels to stay on roofs and other structures for several years without disturbance. The mounting equipment grants the whole system balance and stability.

Utility Meter

Some would say that having a utility meter is optional; but I would say that it is mandatory. If you really want to know exactly how much power you are expending and producing on a regular basis. You can get your hands on cheap utility meters at most hardware stores or you could purchase them online.

Generator

Although the whole purpose of having a fully operational mobile solar power system is to be free from other sources of energy, you still would be wise to have a backup generator on hand. Even if you almost all of your power from your solar power system, there be certain instances in which a backup generator would be useful. If for example you find yourself go through several overcast days without enough sunlight, the generator could temporarily step in to fill that gap.

Also, and perhaps more importantly, if you are out in the middle of nowhere and you suddenly encounter a system failure (it happens) having an emergency generator on board could really save your life. If you do have to resort to using a generator to restore power to your batteries, just be sure not to charge them for too long. Generators can fry batteries if you send too much power to them. To avoid this, actively monitor battery levels as you charge, and make sure you don't charge your batteries more than 95%. Leave at least 5% as a cushion to avoid overcharging.

Chapter 2: Mobile Solar Power Setups

After getting a basic understanding of solar power and the kind of components that make up typical solar power systems, in this chapter let's take it a step further and go over some very specific details and instruction for the kinds of DIY mobile solar power setups you can construct for yourself.

Mobile Mason Solar Powered Light

In order to create your own mason jar solar light, you will need to use a yard based "solar light" that you can stake into the ground. For the purpose of this exercise, go ahead and remove the stake from the light fixture. You can just pull it right out. This will allow you to have the remaining solar panel and the plastic casing for your LED. Next, pick up your solar light and put foam around it.

Now get out a mason jar, and take off the lid. You should just have the ring remaining, leave the ring on the jar. Now put your light through the ring, squeezing it through the center of the ring. With your light in place secure your lid onto the jar. This mobile mason solar powered light is complete

12 Volt Portable Solar Power Bank

This is one of the best mobile solar power projects you could take part in, and one of the most effective. There are quite a few moving parts with this one however, so before we even get started let's just go ahead and list them all right now. For this project you will need:

12 volts to 220-volt inverter
20W Solar Panel
USB Hub
Super Glue
DC to DC USB Booster Circuit
LED leads
Breadboard
N914 diode
Mini slide switch pin

To get started get out your solar panel and solder a pair of solid core wire ends to the negative and positive leads of the solar panel. Now place your solar panel and LED leads into the holes inside your breadboard. As soon as you see the LED lighting come to life you will know that your solar panel is indeed functioning as it is supposed to.

Next you are going to install your DC to DC USB Booster Circuit. Put your DC to DC USB Booster Circuit leads right into the holes of your breadboard, making sure that they are placed within the same row of the breadboard.

Next, you need to install a mini slide switch so that you can always know when your charger is active or not. This is perhaps the most difficult step. But once you locate where your mini slide switch pin needs to go on your breadboard, go ahead and put it in place. Once this is in place, you can then move on to install your N914 diode. This diode will be responsible for controlling the signal in which your current of solar powered electricity will travel to your connected device.

This diode will make sure that the just the right amount of power is released in order to prevent short circuiting your equipment. Put this diode in the same column of your bread board as the DC to DC USB Booster Circuit. Finally, you are going to want to round this little tutorial off by putting your solar panel into your breadboard. Just be sure that your circuits line up and match with what you need on your breadboard and everything should work without a hitch.

Solar Power Mint Tin

This ingenious project turns an everyday mint tin into a solar power house! To get started you will need to acquire for yourself a small mint tin, and an everyday "solar light" the small solar powered lawn accessory that folks drive right into the ground. Once you get your mint tin and your solar light gathered together you can begin. First, get out your mint tin and set it to the side for a moment. Now go to your car and get out your USB car adapter (we all got them, right?) and disassemble it.

Take out the USB port from the adapter. Next, get out your 9-volt battery clip and position it's red wire with the plus side of your USB piece, with the dark wiring up, exposing the minus side. Now solder the two pieces in place with your soldering iron, making sure to hold the solder gun right on the wire until it melts, congeals and hardens. With these wires in place, push your battery into the clip. You will now see the USB light up, indicating that it is active.

Now get that mint container out again, and punch a hole in the side of it. Once you have a hole in your tin, take a pair of pliers and use them to widen the hole. Now push the USB into the hole, before taping it in place. After you have done this affix two solar cells on top of the tin, once the solar cells are in position, run a solar cable from the cells to the USB. You can now start charging your phone from your solar powered mint tin!

Pepsi Can Solar Panel

If you like soda you are going to love this solar power DIY! And don't let the title throw you, it doesn't necessarily have to be Pepsi products that make up your solar panel, you can do it with all manner of soda brands. Because as it turns out, aluminum cans of all variety are perfect conductors of solar energy! For this project you will need about 80 empty cans of soda in all. If you are indeed a major soda aficionado, you should be able to purchase a few 24 pack cases of soda and have just about all the cans you need for this project.

If, however, you do not wish to drink yourself to reach the can quota you could take a more proactive approach and go out and collect them for yourself. I understand that picking cans up off the side of the road is not the most dignified thing that anyone could do, but if you are discreet about it, hardly anyone would notice in the first place. The best way to gather plenty of cans without attracting attention to yourself would be to go to a campground or park early on a Sunday morning. Early Sunday morning hours there shouldn't be too many people out and about, and yet you will no doubt find the whole park littered with the remnants of the day before, including a plethora of aluminum cans.

However, you get a hold of your cans however, just make sure that you clean them out before you use them. The most obvious reason for this would be to make sure they are clean and sanitized before you make contact with them, and the other reason is due to the fact that as much as the surface of an aluminum can is a good conductor, the syrupy soda residue that used soda cans are often coated with—are not. So, your can's can have optimal absorption you will need to thoroughly clean them off. Once your cans are clean, the next thing you are going to want to do is shape them into what will become your solar cells.

Take a pair of wire cutters and use them to cut off both the top and the bottom of the can. Now cut the cans open longwise (be careful not to cut yourself, the edges might be sharp) all the way around. Take a hammer and use it to flatten out your cut open can into one uniform piece of metal. Once your cans are flattened take a drill and start drilling holes in the center of each of your flattened cans. After this, get out some wood working glue and use it to glue all of your cans to the interior of your solar panel.

Once your glue has dried, you can either solder your wires direct to the apparatus or in a real pinch you can get by with just hooking up a pair of car jumper cables. If using the latter, simply hook the jumper cables up to where they are touching the solar cells and then hook the other end up to your battery run through an inverter. Once you do you will have fully function solar mobile power.

Chapter 3: Monitoring Your Power Bank Voltage

It can't be stressed enough how important proper monitoring and record keeping is when it comes to mobile solar power systems. This chapter runs through all of the ways that you should keep tabs on your solar power.

Calculate Just How Much Power You Need

The first step of calculating the amount of power you might need, is to take an inventory of all of the devices and appliances that you would like to run on off of your solar power system. Once this calculation has been made you need to choose which devices you would like to use and for how many hours you would like to use them. It needs to be calculated in hours so that you can record their "wat hours". If for example, you would like to run a 10 W CFL bulb for 4 hours. You could run this snazzy little equation:

Watt Hour = 10 W x 4 hours = 40-watt hours

Know How to Log and Maintain Data

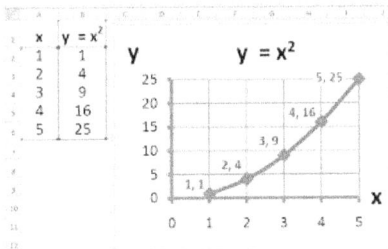

It stands to reason that it is very important to keep a coherent record of all of the data in regard to overall output and input on all of your solar powered systems. You should keep a special database on one of your computer's just for this task. Microsoft Excel spreadsheets are perfect for such record keeping. But if you have to, you could always keep a journal by hand—it just might be a little bit more tedious! Nevertheless, be sure you know how to log and maintain your data!

Don't Fry Your Battery

The potential capacity of a battery is noted by its overall charge. In order to not overload its storing capacity, you need to always make sure that you don't leave them connected to your solar panels or generators for two long. As a rule, you should never charge your batteries more than 95% percent, that way you still have a little bit of room in the battery before you reach the absolute threshold of its storage capacity. Always be sure to leave this space in your battery to you don't fry it!

Chapter 4: Further Maintenance

Everything needs maintained. Cars need oil changes, grass needs to be cut, dogs need to be walked, and solar panels need to have regular maintenance.

Regularly Inspect Your Panels

Solar panels are out in the open, and exposed to the elements. Having that said, there are many things that could adversely affect them from the environment. Even something as simple (and gross) as a bird defecating on top of the unit could wreak havoc. In order to counteract such contingencies, you will need to regularly inspect your panels. Hail storms and wind and rain damage are one of the most common problems that most outdoor solar panel systems face, so if the weather has been especially bad in your area, you might want to climb up on a ladder and take a look.

Keep Your Solar Panels Clean

Sometimes solar panels become completely covered in dust and pollen from the environment. If this happens it can have a significant impact upon the productivity of your mobile solar power unit. In order to prevent such things from occurring you may need to regularly clean your solar panels themselves. Clean them with just plain water, don't try to use any special solvents. To make things even easier, some are able to even use their garden hose to spray the surface of their solar panels clean.

Keep Track of Sun and Shade

In order to keep your solar panels at their most effective, you are going to have to be able to track just how much sun and shade are prevalent at any given time. This means that you need to have your solar powers placed in the best possible position in order to take full advantage of the current weather conditions. This is no doubt of prime focus when a solar power system is first installed, but keep in mind that these conditions can change over time.

Nearby buildings or other structures may be erected that create shade and block out the sunlight that reaches your unit. Vegetation may also grow up around the solar panel over the years, that creates shade and limits productivity as well. These things were not in place during your first install, but over time such new factors may come into play, so make sure you keep track of overall sun and shade.

Monitor Overall Usage

Although it is tempting to believe that our usage levels of electricity can be kept under wraps without much oversight, this is usually not the case. You need to be able to monitor your overall input and output on a regular basis. This is especially the case if you end up with a surplus of power and would like to sell it back to utility companies. For these companies to purchase your solar power you need to have an accurate usage history to give to them.

Fortunately for us, most inverters—a device that is just about mandatory anyway—have built in meters that allow you to directly gauge your overall daily power output. In many instances you can even download associated apps direct to your smart phone so that the inverter can send them to your mobile device no matter where you are. Be sure to use these new innovations to monitor your overall usage of power.

Conclusion: Mobile Power on the Go!

In the modern world we just can't get by without some form of power to charge our phones, laptops and a whole host of other devices. This book seeks to provide solar power solutions to that energy need. Even if you find yourself lost in the middle of the wilderness, follow the instructions presented in this guide and you will have mobile power on the go—wherever you go! Thank you for reading!

www.ingramcontent.com/pod-product-compliance
Lightning Source LLC
Chambersburg PA
CBHW071000220526
45471CB00007B/3118